Algebra

100 Fully Solved Equations To Explain Everything You Need To Know To Master Algebra!

By Math Wizo

First, Get My FREE GIFT!

For a limited time, you can download this book for FREE!
Get it by going to: https://go.mathwizo.com/1

ISBN: 9781792889660

TABLE OF CONTENTS

ALGEBRA:
EXPLAINED THE EASY WAY WITH 100 FULLY SOLVED EQUATIONS

You are about to embark on a journey into mathematics, so don't freak out. More specifically, this journey is all about algebra, which is basically one of the major divisions of math. While your experience with algebra may have so far been confusing, this book is designed to make it simpler for you. Algebra is, in fact, a way to look at math problems that actually makes math problems simpler. This is especially true of word problems, which can be converted easily into basic algebra problems that you will be able to solve after reading this.

The key thing to know about algebra is that is based not only on numbers but on things called "variables". What is a variable? A variable is something about the problem that you don't know the value of. In fact, algebra can be considered a type of mathematics that is concerned with the solving or determination of what number value the variable has. In algebra, mathematicians use placeholders, such as "x", "y", and "z" to indicate that these represent numbers of which we do not yet know the value. Your job is to know what the value of the variable is based on the rest of the known numbers in the equation.

The rest of algebra is basically the same arithmetic you've known since grade school. Addition is still addition, subtraction is still subtraction, and so on. Algebra also uses things like the square of something and the square root of something, which may not be something you truly understood in middle school or high school. The square of something is nothing more than the number multiplied by itself. This would be written like this: Two squared or 2^2 is the same as saying 2 x 2 or 4. It is written as $2^2 = 4$. In the same way, the cube of something is the number multiplied by itself three times. $2^3 = 2$ x 2 x 2 = 4. The square root is the reverse of this. The square root or $\sqrt{}$ of something is the reverse answer to the square of something. For example, the square root of 4 or $\sqrt{4}$ is 2 because 2 x 2 = 4.

The other thing you'll have to remember and do as part of algebra involves the idea that equations are always balanced. In other words, the equation 2 + 2 = 4 means that everything on the left side of the equation or "=" sign is the same as everything on the right side of the equation. And, if you'll think back to grade school, you can manipulate these equations. You can take 2 + 2 = 4 and turn it into 2 = 4 – 2. Easy, right? Well, in algebra, you'll see things like x + 2 = 4, and you'll have to "solve for x", meaning that you have to put x on one side of the equation and figure out what x means. It looks like this: x = 4 – 2. The answer is, of course, "2". This is algebra. While there are harder problems than this, you can use these basic rules to solve equations that, in this book, will get harder as we progress.

To be clear about the nomenclature, we will use x, y, z, a, b, and c to represent the variables and it will be up to you to determine the number the variable represents. There will also be some differences in how things are represented in print. In order to avoid confusion, we'll use these symbols to represent mathematical equations:

- **Addition.** This will always involve the "+" sign you are used to seeing.
- **Subtraction.** This will always involve the "–" sign you are used to seeing.
- **Multiplication.** This will be written in one of four ways. The "*" sign will be used in some equations, such as 2 * 6 = 12. More often, however, you will see it this way: 2(6) = 12 with the parentheses used to represent multiplication with the number next to it. It may also be written like this: (2)(6) = 12. This also represents 2 multiplied by 6. This allows for equations like this: 2(2+4) = 12. In this situation, the mathematical protocol is to do what's in the parenthesis first and then do the mathematical problem of multiplication next. Why do this? Because 2 * 2 + 4 is completely ambiguous. 2 * 2 = 4. Add 4 to that and you get 8. BUT, 2 * (2+4) is the same as 2 * 6 and this answer is 12, not 8. The parenthesis tells you which set of operators to do first. Finally, you might see 2x or 3y. The combination of the integer and the variable together like this means you are supposed to multiply them.
- **Division.** This is not represented by the ÷ sign (although it COULD be used). Instead the "/" symbol is used so that 2/4 = two-fourths or one-half. In the same way 4/(2+2) is 4/4 or 1. In algebra, you might

see 2/x or (3 -2)/y. This is no different than you see when you're just using integers.

- **Squared or cubed.** This will be as you know it. Squared is x^2 and cubed is x^3 and so on. Know that it is possible to have any number other than 2 or 3, such as x^n in which n represents any possible number.
- **Square root.** The symbol "√" will be used to represent the square root of something. It will be possible to have the equation √(2+2), in which the entirety of 2 + 2 is under the square root. Again, the mathematical operators inside the equation are done first and then the square root.

So, in this book, we will look at easy, medium, and hard questions involving algebraic equations. Not only will we solve them together but we will work together to see how these equations get solved. So, if you're ready, we'll get started.

CHAPTER 1:
THE BASICS OF SINGLE VARIABLE PROBLEMS

In this chapter, you'll get your feet wet and will learn how mathematical operators work in solving simple algebraic equations. We won't do many word problems; there will be just a few so you can see how algebra can help.

- **Problem 1.** Solve $25(x + 2) = 250$
  In this case, as in all cases of algebra, you will be asked to solve for x. You first need to get as many things on the side of the equation opposite the variable, which in this case is x. Start by putting 25 on the opposite side of the equation. This looks like this: $(x + 2) = 250/25$. In order to move the 25, you need to put it on the opposite side and divide it into 250.

  Another way to do this is to divide both sides by 25. So, $25(x + 2)/25 = 250/25$. That's safe to do because you can divide, multiply, add or subtract anything so long as the same thing is done to both sides of the equation. Either way, you get $x + 2 = 10$ because $250/25 = 10$. This gets a lot simpler. The next thing to do is to move the 2 to the other side of the equation to get $x = 10 - 2$. For this reason, the answer is $x = 8$. That's all it takes. Whew!

- **Problem 2.** Solve $(x - 4)/5 + 15 = 19$.
  Again, get everything, step-by-step onto the opposite side of the equation as the variable and you've solved for the variable. Start with the 15. If it's added on the left side, subtract it from both sides to move it. Then you get $(x - 4)/5 = 19 - 15$. This translates into $(x - 4)/5 = 4$. Now, we have to get rid of the 5 on the left-hand side. The operators within the parenthesis stay together until the very end. This leaves you with $(x - 4) = 4 * 5$. Again, if it's divided on the left side, you have to multiply it on both sides to balance the equation.

This leaves you with $x - 4 = 20$. Now, add the 4 to both sides of the equation because it's subtracted on the left side to get $x = 4 + 20$. The answer? $x = 24$

- **Problem 3.** Solve $500 = 25(x - 5)$.
  We'll do this a couple of ways. First, you can do it this way: Start by switching the equation around to be $25(x - 5) = 500$. Then move the 25 to make $(x - 5) = 500/25$ or $x - 5 = 20$. Then move the 5 to make $x = 20 + 5$ or $x = 25$. You can also do it this way: Start with $25(x - 5)$. You can multiply both the x and the 5 by 25 to get $25x - 25(5) = 500$. This leads to $25x = 500 + 25(5)$ or $25x = 500 + 125$. This becomes $25x = 625$. Now you have to divide both sides by 25 to have x isolated to the left side of the equation to make $x = 625/25$. Using a calculator, you get $x = 25$.

Sometimes, the first way is easier and other times, the second way is easier. Just know that you can solve this equation either way. As always, when you move something from one side of the equation to the other, addition on one side translates to subtraction on the other side and multiplication on one side translates to division on the other side. We haven't had to do it yet but know that squaring something on one side of the equation requires getting the square root on the opposite side of the equation.

- **Problem 4.** Solve $(x^2 + 20)/2 = 60$.
  This equation will get into the issue of squares and square roots. First separate the 2 from the left side of the equation by multiplying it on the other side. This makes $x^2 + 20 = 120$. Then move the 20 to the right side by subtracting it from both sides to make $x^2 = 120 - 20$ or $x^2 = 100$. You will need a calculator or your good memory to get $x = \sqrt{100}$. The square root of 100 is 10. This leads to the answer, which is $x = 10$. It's a little more complicated but you can also solve the equation this way: $(x^2 + 20)/2 = 60$, which becomes $x^2/2 + 20/2 = 60$ or $x^2/2 + 10 = 60$.

You can make this less complicated by subtracting 10 from both sides to make $x^2/2 + 10 - 10 = 60 - 10$. Cross out the 10s on the left side and combine the numbers on the right side to get $x^2/2 = 50$.

Now multiply 2 on both sides to get $x^2/2 * 2 = 50 * 2$. Again, cross out the 2s on the left side and multiply on the right side to get $x^2 = 100$. Now take the square root of both sides to get $\sqrt{x^2} = \sqrt{100}$. Again, the same operation is done to both sides to keep things equal. The square root of a squared number, even a variable squared is the number or variable itself. This leads to x =10.

- **Problem 5.** Solve $200 - 4x = 20$.
  This involves negative numbers but it needn't be too challenging if you remember a few rules. Start with moving the 200 to the right side of the equation to make $-4x = 20 - 200$. This takes +200 on the left side and makes it -200 on the right side. the equation becomes $-4x = -180$. The good news is that, with this question, you can simply get rid of the negatives on both sides and turn it into $4x = 180$. It's basically the same equation only simpler to maneuver. Then divide the 4 on the other side to get $x = 180/4$ or $x = 45$. Not so sure we could do that? Leave the $-4x = -180$ and divide the -180 on the right side with -4. In relatively simple mathematics, you get $x = -180/-4$, which still makes x = 45. When you divide two negative numbers by each other, you get a positive number. See, you can make the negatives disappear as long as you do the same thing on both sides.

- **Problem 6.** Solve $\sqrt{y}(4) = 20$.
  Since the square root part is the hardest to solve, divide both sides of the equation by 4 to get $\sqrt{y} = 20/4$ or $\sqrt{x} = 5$. Since the opposite of square root is to square something on the opposite side to get $y = 5^2$. This is the same as $y = 5 * 5$ or $y = 25$.

- **Problem 7.** Solve 3x + 5x = 16

  First of all, do not panic and feel like you have 2 "x" values to determine. This is the beauty of variables. The "x" in the equation is the same value in both instances. You simply need to manipulate the equation to have just one "x". When things are "together" such as 3x or 5x, you add the two terms together. For example, what if x was a number like 3? You'd get 3(3) + 5(3). You then get 9 + 15 and you can add them together to get the correct number. It is no different in algebra; 3x + 5x is the same as 8x. This leads to 8x = 16. Now just divide by 8 on both sides of the equation in order to isolate the x. This leads to x = 16/8 or x = 2.

- **Problem 8.** Solve 2(3x + 4x) = 42

  This equation can be solved two ways. First, you perform the operation inside the parentheses in the equation to get 2(7x) =42 by adding 3x and 4x. This leads to 14x = 42. Divide both sides by 14 to get x = 42/14 or x = 3. You can also solve the equation this way: multiply each of the variable expressions by 2 in the parentheses to get:

  $$2(3x) + 2(4x) = 42.$$

  Then you get 6x + 8x = 42. This naturally leads to 14x = 42 or x = 3. The answer is the same and you can do it either way. Just know that the answer will be the same.

- **Problem 9.** Solve (25x -20x)/5 = 5x – 4

  Again, this is not a cause for panic because there is more than one x. They both have the same value. The goal is to get all of the "x" values on one side of the equation and all the numbers on the other side. For simplicity sake, start with the operations inside the parentheses and make 5x/5 = 5x – 4. Now divide the left side of the equation to get x = 5x – 4. Now, all the x values need to be on the same side of the equation. Take the +5x (a positive number) and make it a -5x (a negative number) and put it on the left side of the equation by subtracting it from both sides. This leaves x – 5x = -4 or -4x = -4. Divide both sides by -4 (the opposite of multiplication that's on the left side of the equation) to get this:

  $$x = -4/-4$$
  $$x = 1$$

  This looked complicated but, in the end, it wasn't so hard after all. Let's do another one like this so you feel more comfortable doing it.

- **Problem 10.** Solve [2x + (4x+ 4x)/2] = 24

  This starts with doing the operation inside the inner parentheses. This leads to this:

  $$2x + 8x/2 = 24$$

  This leads to 2x + 4x = 24

  Combine the two variable terms to get 6x = 24

  Divide both sides by 6: x = 4

  Hopefully, things are getting easier for you to comprehend, even though there is more than one variable term. We will do a few more of these manipulations and then go on to some single variable word problems.

- **Problem 11.** Solve 16(10-3x) = 32x

  This time, you cannot start with the parentheses because the parentheses do not contain like terms, so multiply each of the terms in the parentheses by 16 to get this:

  $$16(10) - 16(3x) = 32x$$
  $$160 - 48x = 32x$$

  Now get the x values on the same side of the equation: 160 = 32x + 48x

  Combine the variable terms: 160 = 80x

  Divide the both sides by 80: 160/80 = x

  The answer: x = 2

- **Problem 12.** Solve [(8x + 2x)/(6 -x)] = 10

  Start by doing what you can inside the parentheses. This leaves you with 10x/(6-x) = 10

  This is not that hard to do. Take the parentheses (6-x) and multiply it on both sides (to negate the division on the left side) to make this:

  $$10x = 10(6-x)$$

  Distribute the right side: 10x = 60 − 10x

  Move all the variable terms to the left side: 10x + 10x = 60

  $$20x = 60$$
  $$x = 60/20$$
  $$x = 3$$

- **Problem 13.** Solve (24x – 16x)/(4-x) = 4x.
  Of course, this involves a lot of variable terms but, in reality, IT'S ALL THE SAME VARIABLE so relax! Let's tackle it first by doing the operations inside the first parentheses: 24x – 16x is 8x. This gives us this to continue with:

$$8x/(4-x) = 4x$$

Now multiply both sides by the expression inside n the parentheses and distribute on the right side:

$$8x = 4x(4-x)$$
$$8x = 16x - 4x^2$$
$$8x -16x = -4x^2$$
$$-8x = -4x^2$$

Because both sides are negative, you can "equalize" them and make them both positive: $8x = 4x^2$

Divide the left side by 4 to get: $2x = x^2$

This is where it gets a little tricky. The correct way to solve this equation is to set it equal to zero because it is a quadratic equation. This gives you $x^2 - 2x = 0$. Then since both terms have an x in them, factor the x out as a greatest common factor, resulting in $x(x - 2) = 0$. Now the Zero Product Property states that if the product of terms gives you zero, then one or all of the terms have to be zero. This gives $x = 0$ or $x - 2 = 0$, which results in two solutions to the equation.

$$x = 0 \text{ or } x = 2$$

- **Problem 14.** Solve 2x/(4-x) = -2x.
  You can't solve the number and variable in the parentheses because it's in the denominator, so move them to the other side by multiplying both sides and distributing:
  $$2x = -2x(4-x)$$
  $$2x = -8x - (-2x^2)$$
  You can get rid of the double negatives by making it positive: 2x = -8x + 2x^2
  This gives you a quadratic equation, so set the equation equal to zero.
  $$0 = 2x^2 - 10x$$
  Since both terms have a $2x$ in them, factor the $2x$ out as a greatest common factor, resulting in
  $$0 = 2x(x-5)$$
  Now use the Zero Product Property to solve the equation, so $2x = 0$ or $x - 5 = 0$.
  The solution to the equation is $x = 0$ or $x = 5$.

- **Problem 15.** Solve -4(x + 2)/(x − 4) = -16
  Start by multiplying the (x − 4) on both sides to get this:
  $$-4(x + 2) = -16(x - 4)$$
  Now multiply the numbers outside of the parentheses with what's in the parentheses using distribution:
  $$-4x - 8 = -16x - (-16 * 4)$$
  $$-4x - 8 = -16x + 64$$
  Now move the variables and the numbers to opposite sides:
  $$-4x + 16x = 64 + 8$$
  $$12x = 72$$
  Divide both sides by 12 to get x = 72/12
  $$x = 6$$

- **Problem 16.** Use algebra to solve this problem. Joe has a rectangular swimming pool with one side x feet long and the other side twice as long as the first side. He knows that the sum of the lengths of all of the sides of the pool is 120 feet. How do you solve for x, which is the length of the two opposite sides?

One side is x and the other side is 2x.

The lengths of all the sides are added up to make x + x + 2x + 2x = 120.

combine the terms on the left side: 6x = 120

Divide both sides by 6: x = 120/6

x or the length of one side is 20 feet. The length of the other side is 40 feet.

- **Problem 17.** Use algebra to solve this problem. Joe wants to know the square footage of another pool that has a side that is x feet long and another side that is 3x feet long. He determines that the distance around the pool is 160 feet. What is the area of the pool?

This actually involves two equations and two things you don't know:

The first equation is this: x + x + 3x + 3x = 160

The second equation is x(3x) = y, where y is the area.

Solve the first equation: 8x = 160; x = 160/8 or x = 20

Now solve the second equation: (20)(3)(20) = y; y = 1,200 square feet

- **Problem 18.** Use algebra to solve this problem. Mary has a round pool with an area of 1200 square feet. She has a square tarp. How big of a tarp does she need to cover the pool?

  To solve this problem, you need to know the diameter of the round pool which is 2r or two times the radius. The edges of the tarp need to be at least as long as the diameter of the pool, so you need the diameter of the pool. This means you have to solve the equation for the area of the pool: πr^2 = area. The value of π is about 3.14 (it's actually a very long number that doesn't make much sense to use because it isn't practical). Let's solve the equation:

  $$\pi r^2 = 1200$$
  $$r^2 = 1200/\pi$$
  $$r = \sqrt{1200/\pi}$$
  $$r = 19.55 \text{ feet}$$

  The diameter of the pool is twice the radius so you need a square tarp that is about 39.10 feet by 39.10 feet in total dimensions.

- **Problem 19.** Use algebra to solve this more challenging problem. Amy has a box that has a side that is x inches, another side that is twice as long, and a depth that is also twice as long; the total volume is 500 inches. How many balls with a diameter of 5 inches can be placed into a box of this space? To solve this, you first need to know what the sides of the box are. The volume will be the length times the width times the depth or x * 2x * 2x = $4x^3$. The equation is this:

  $$4x^3 = 500$$
  $$x^3 = 500/4 = 125$$

  The cubic root of 125 or the value of x is 5 because 5^3 is 125. Because a ball is round, the only number that counts is the length of the shortest side, which is 5 inches. Imagine a rectangular box that is 5 inches wide by 10 inches long by 10 inches high. If the balls have a five-inch diameter, 2 layers of 2 balls can fit into the box. The answer is 4 balls.

- **Problem 20.** Use algebra to solve this problem. Jane ran laps around a track for a total of x miles, while Jeff ran twice as far and Julie ran 4 times as far as Jeff and Jane together. Together they ran 45 miles. How many miles did each person run?

   Just think about the equation first: x + 2x + 4(x + 2x) = 45 miles

   Now solve the problem: x + 2x + 4(3x) = 45

   $$3x + 12x = 45$$
   $$15x = 45$$
   $$x = 45/15$$
   $$x = 3$$

This means that Jane ran 3 miles, Jeff ran 6 miles, and Julie ran 36 miles.

- **Problem 21.** Solve this problem using algebra. In a bike ride in which cyclists raised $5 per mile, person A rode x miles, person B rode 4 times that many, and person C rode half as many miles as person A and person B together. The goal was to raise $1000 dollars each. Together they rode for 825 miles. Did any person raise at least $1000 in the ride?

   The equation is this: x + 4x + (x + 4x)/2 =825 miles

   $$5x + x/2 + 4x/2 = 825$$
   $$5x + x/2 + 2x = 825$$
   $$7x + ½ x = 825$$
   $$7.5x = 825$$
   $$x = 110$$

This means that person A raised $550, person B raised $2,200, and person C raised (110 + 440)/2 * 5 dollars or $1375. This means that person B and person C made their $1000 goal.

- **Problem 22.** Solve this word problem using algebra. You have a trapezoid with two equal slanted sides with a length of x, a long side of 3x and a short side of 2x. The shape looks like this:

trapezoid

The perimeter of the trapezoid is 28. Solve for x.
This is actually easy to solve. Add x + x + 3x + 2x, which must add to 28. The equation is this:

$$x + x + 3x + 2x = 28$$
$$7x = 28$$
$$x = 28/7$$
$$x = 4$$

- **Problem 23.** Use algebra to solve this problem. You have a circle with an area of 60 cm^2 and you want to draw a circle that has 4 times the diameter. What is the diameter of the circle you are trying to draw?

   The area of a circle is πr^2 cm^2, where r is the radius of the circle. You know that the diameter of the existing circle is 2r. (The diameter is twice the radius).

   You know that the area of the existing circle is 60. Find the diameter of the existing circle.

   $\pi r^2 = 60$; $r^2 = \frac{60}{\pi}$; $r^2 = 19.108$; $r = \sqrt{19.108}$; $r = 4.37$, so the diameter of the existing circle is 8.74.

   Since you want the diameter of the new circle to be 4 times the diameter of the existing circle, multiply the diameter of the existing circle by 4:

   $$d = 4 \cdot 8.74 = 34.96$$

   The diameter of the circle you are trying to draw is 34.96 centimeters. Okay, I don't know about you, but I think that was a hard one. Still, there was only one variable.

- **Problem 24.** Use algebra to solve this word problem. There are 3 runners. Runner A ran x miles. Runner B ran three times as many miles. Runner C ran 3 miles less than three times the number of miles ran by runners A and B together. The total number of miles ran was 61. How many miles did each runner run?

   We know that x + 3x + 3(x + 3x) – 3 = 61. Solve for x.

   $$x + 3x + 3x + 9x - 3 = 61$$
   $$16x - 3 = 61$$
   $$16x = 61 + 3$$
   $$16x = 64$$
   $$x = 4$$

   Runner A ran 4 miles, Runner B ran 12 miles, and Runner C ran 3(4 + 12) – 3 miles or 45 miles.

- **Problem 25.** Use algebra to solve this problem. You have three children who collect marbles. Child A collected x marbles, child B collected 5 fewer marbles than child A, and child C collected twice as many marbles as child B. The total number of marbles collected was 245. How many marbles were collected by each child?

Solve this problem: $x + x - 5 + 2(x - 5) = 245$:

$$2x - 5 + 2x - 10 = 245$$
$$4x - 15 = 245$$
$$4x = 245 + 15$$
$$4x = 260$$
$$x = 65$$

Child A collected 65 marbles, child B collected 60 marbles, and child C collected 120 marbles.

CHAPTER 2:
EXPLORING SIMPLE TWO VARIABLE PROBLEMS AND FACTORING POLYNOMIAL TERMS

This chapter will expand on algebra to include problems that involve two variables and polynomial terms. Let's start with two variable equations. When there are two variables, there will be two separate equations, called a system of equations that need to be solved with an incorporation of the two equations into one another to solve the equations together. Just remember that, in order to solve two variable equations, you need two equations in order to be able to solve for both equations.

- **Problem 26.** For these first few questions, we'll simply manipulate two variable problems in order to see how this works. These will look more complex than the problems we've had in the past; however, they are worked like any other problem. Technically, they aren't solvable but give you a sense of the "relationship" between two variables. Start with this one:

    Solve $2y + 5 = (4x + 2)/2$ for y. In this equation, you can "solve" for either variable but we'll solve for "y" in this equation:

$$2y + 5 = (4x + 2)/2$$
Move the 5 to get this: $2y = (4x + 2)/2 - 5$
Clean this up a little bit: $2y = 4x/2 + 2/2 - 5$
$$2y = 2x + 1 - 5$$
$$2y = 2x - 4$$
Move the 2 on the left by dividing by 2 on both sides: $y = 2x/2 - 4/2$
Solving for y you get: $y = x - 2$

    This will give you the relationship between x and y but does not actually solve for these values. This will be important when you have two equations to solve.

- **Problem 27.** Let's manipulate another problem before getting into the 2-equation situation. Solve this problem for x. $(4y + 4)/2 = (6x - 12)/6$. This time, we will solve for x:

Clean up both sides. You can't do the operators in the parentheses because they are a combination of numbers and variables.
$$4y/2 + 4/2 = 6x/6 - 12/6$$
$$2y + 2 = x - 2 \text{ (This just "cleans up" both sides of the equation).}$$
Move the 2 on the right to the left side: $2y + 2 + 2 = x$
$$x = 2y + 4$$
Why is this manipulation so important? Because, when you know the relationship between the two variables, you can solve for both variables if you know the value of one of the variables.

- **Problem 28.** Solve this problem first for x: $(4x - 10)/2 = 2y + 4$.
Multiple by 2 on both sides of the equation to get this: $4x - 10 = 2(2y + 4)$
Clean up the equation: $4x - 10 = 4y + 8$
$$4x = 4y + 8 + 10$$
$$4x = 4y + 18$$
Divide by 4: $x = 4y/4 + 18/4$
$$x = y + 9/2$$
This is as far as you can go unless you have more information. What if you know y = 8? Then you can solve for x this way: $x = 8 + 9/2 \ x = 25/2$. Solving for x and y, you get $x = 25/2$ and $y = 8$.

- **Problem 29.** Now, things will appear more challenging but will actually be completely solvable. Let's try solving these equations: $(9x + 6)/3 = 12y - 4$ and $3y + 6 = 2x$. These are two-variable problems in which both x and y need solving: Start by solving for y in the first equation. Technically, you can start anywhere and solve for either variable first.

$$(9x + 6)/3 = 12y - 4$$
$$9x/3 + 6/3 = 12y - 4$$
$$3x + 2 = 12y - 4$$
$$3x + 2 + 4 = 12y$$
$$3x + 6 = 12y$$

Solve for y: $y = 3/12 * x + 6/12$
$$y = x/4 + 1/2$$

Now you know the relationship between x and y even though you don't know the actual values of x and y. Now change the second problem to look like this:

$$3y + 6 = 2x$$
$$3(x/4 + 1/2) + 6 = 2x$$

This just puts the y variable relationship inserted into the second equation. This leads to this:

$$3x/4 + 3/2 + 6 = 2x$$
$$[(¾x + 3/2) + 6]/2 = x$$
$$3x/(4 * 2) + 3/(2 * 2) + 6/2 = x$$
$$(3/8) x + ¾ + 3 = x$$

Now move the x variables to one side of the equation:

$$(3/8)x - x = -3/4 - 3$$
$$(3/8)x - (8/8)x = -3/4 - 3$$
$$-5/8x = -15/4$$

Multiply both sides of the equation by -8/5. This is the equivalent of dividing by -5/8 on both sides:

$$x = (-15/4)(-8/5)$$
$$x = 6$$

Now solve for y: Since $y = x/4 + 1/2$
Insert the value of x into the equation: $y = 6/4 + 1/2$
$$2$$
The answer leads to x = 6 and y = 2

- **Problem 30.** Let's do a few more so you can get comfortable with it:
Solve for x and y. 2x/(4x + 6) – 6 = y – 4 and 4y = ½(x) – 13
Start by solving for x on the first equation (again, you can solve for x
and y in either equation and you will get the same answer):

$$2x/(4x + 6) - 6 = y - 4$$

Start by solving the first equation for y.

$$\frac{2x}{4x + 6} - 6 = y - 4$$

$$\frac{2x}{4x + 6} - 2 = y$$

Multiply both sides by 4.

$$4\left(\frac{2x}{4x + 6} - 2\right) = 4y$$

Simplify the left side

$$\frac{4x}{2x + 3} - 8 = 4y$$

Now since both equation equal 4y, set both equations equal to each
other.

$$\frac{4x}{2x + 3} - 8 = \frac{x}{2} - 13$$

Multiply the entire equation by $2x + 3$.

$$\left[\frac{4x}{2x + 3} - 8 = \frac{x}{2} - 13\right](2x + 3)$$

$$\frac{4x}{2x + 3} \cdot (2x + 3) - 8(2x + 3) = \frac{x}{2}(2x + 3) - 13(2x + 3)$$

$$4x - 16x - 24 = x^2 + \frac{3x}{2} - 26x - 39$$

$$0 = x^2 + \frac{3x}{2} - 26x - 39 - 4x + 16x + 24$$

$$0 = x^2 - 12.5x - 15$$

Solve this equation using the quadratic formula, to the nearest
tenth. In this equation a = 1, b = - 12.5, and c = - 15.

$$x = \frac{12.5 \pm \sqrt{(-12.5)^2 - 4(1)(-15)}}{2 \cdot 1}$$

$$x = \frac{12.5 \pm \sqrt{156.25 + 60}}{2}$$

$$x = \frac{12.5 \pm \sqrt{216.25}}{2}$$

$$x = \frac{12.5 \pm 14.7}{2}$$

$$x = \frac{12.5+14.7}{2} \text{ and } x = \frac{12.5-14.7}{2}$$
$$x = 13.6 \text{ and } x = -1.1$$

- **Problem 31.** A few more problems until we get more complex: Solve these equations for x and y:

$$(4y + x)/4 = 2y \text{ and } 4x + 2y = 24$$

It's simpler to solve for y in the second equation:

$$4x + 2y = 24$$
$$2y = 24 - 4x$$
$$y = 24/2 - 4x/2$$
$$y = 12 - 2x$$

Now that this equation is simplified, insert it into the first equation:

$$(4y + x)/4 = 2y$$

Start by doing this to simplify the equation: 4y + x = (4)(2y)

$$4y + x = 8y$$

Now insert the solution for y: 4(12 − 2x) + x = 8(12 − 2x)

Simplify the equation: 48 − 8x + x = 96 − 16x

Put all the x variables on the left and all the numbers on the right:

$$-8x + x + 16x = 96-48$$

Simplify: 9x = 48

$$x = 48/9 \text{ or } x = 16/3$$

Place this value into y = 12 − 2x (this is the simplified second equation).

$$y = 12 - 2(16/3)$$

y = 12 − 32/3 or y = 36/3 − 32/3 or y = 4/3

The answers are x = 16/3 and y = 4/3

- **Problem 32.** Let's try something a little different by solving these two equations:

$$2x + y + 2 = 4x \text{ and } 2y + 6 = 6x$$

This really isn't so hard. Solve for y in the first equation:

$$2x + y + 2 = 4x$$
$$y = 4x - 2x - 2$$
$$y = 2x - 2$$

Now solve the second equation, substituting what you know about y:

$$2y + 6 = 6x$$
$$2(2x - 2) + 6 = 6x$$
$$4x - 4 + 6 = 6x$$
$$4x + 2 = 6x$$
$$2 = 6x - 4x$$
$$2 = 2x$$

Divide both sides by 2 to get x = 1

Now plug this answer into the first simplified equation y = 2x − 2

$$y = 2(1) - 2$$
$$y = 2 - 2 \text{ or } y = 0$$

The solution is x = 1 and y = 0

- **Problem 33.** Solve this problem: $6y + (2y - 4)/2 = 12x - 6$ and $x + 2y = 20$.

Start by solving for x in the second equation:

$$x = 20 - 2y$$

Now use this to solve the first equation:

$$6y + (2y - 4)/2 = 12x - 6$$
$$6y + 2y/2 - 4/2 = 12(20 - 2y) - 6$$
$$6y + y - 2 = 240 - 24y - 6$$

Put the variables on one side and the numbers on the other side:

$$6y + y + 24y = 240 - 6 + 2$$
$$31y = 236$$
$$y = 236/31$$

Place this number into the simplified second problem:

$$x = 20 - 2y$$
$$x = 20 - 2(236/31) \text{ or } x = 20 - 472/31$$

Then multiply 20 by 31/31ths to get x = 620/31 − 472/31

$$x = 148/31$$
The answers are $x = 148/31$ and $y = 236/31$

- **Problem 34.** We'll do two more and then think about some word problems. Solve this algebra problem:

$$(2x + 8)/4 + 4 = (y - 4)/2 \text{ and } y + x = 4 + 2x$$

Start by solving the second equation for y:

$$y + x = 4 + 2x$$
$$y = 4 + 2x - x$$
$$y = 4 + x$$

Now solve the first equation by inserting that answer into the first equation:

$$(2x + 8)/4 + 4 = (y - 4)/2$$
$$(2x + 8)/4 + 4 = (4 + x - 4)/2$$
$$2x/4 + 8/4 + 4 = x/2$$

Simplify this: $x/2 + 6 = x/2$

Put all the variables on one side and all the numbers on the other side, reversing the pluses and minuses as you do this:

$$x/2 - x/2 = -6$$
$$0 = -6$$

The variable cancels leaving us with a false statement because 0 does not equal -6. Therefore, this problem has no solution.

- **Problem 35.** From what you've learned so far, you can see that any two equations with two variables can be solved with an answer. We haven't gotten into more complicated questions like $xy = 9$; this will be the topic of another chapter. We'll do a few more problems on factoring after this problem and then do some two-variable word problems. Solve this problem:

$$(3y + 2x)/2 + 7 = y \text{ and } 3y - x = 2$$

Start by simplifying the second equation by solving for x:

$$3y - x = 2$$
$$-x = 2 - 3y$$

Reverse the pluses and minuses: $x = 3y - 2$

Now solve the first equation by substituting this into the first equation:

$$(3y + 2x)/2 + 7 = y$$
$$3y/2 + x + 7 = y$$

Insert the first equation: $3y/2 + 3y - 2 + 7 = y$

Put all the variables on the left and all the numbers on the right (reverse the pluses and minuses):

$$3y/2 + 3y - y = -7 + 2$$
$$3y/2 + 2y = -5$$
$$7y/2 = -5$$

Multiply both sides by 2/7 to make $y = -5(2/7)$ or $y = -10/7$

Plug this into the simplified first equation to get:

$$x = 3y - 2$$
$$x = 3(-10/7) - 2$$
$$x = -30/7 - 2 \text{ or } x = -44/7$$

The answers are $x = -44/7$ and $y = -10/7$.

Factoring Algebraic Terms

The next part of this chapter involves factoring polynomial terms. In this section, you'll have to know what an algebraic polynomial term is. This is an algebraic term that involves any combination of multiplication, division, subtraction, and addition of a number, a variable, and any variable that has a positive exponent. It doesn't involve negative exponents, such as x^{-5}.

Factoring is dividing a number or equation into its component parts. If you'll remember factoring in arithmetic, you need to take a number like 96 and factor it into the numbers that multiply to make it: 2 x 6 x 8 or 2 x 2 x 3 x 2 x 2 x 2. There won't be this number of factors in most of the problems you'll see in this part of algebra but you will see some interesting factoring. You will have to factor things like this $x^2 + 2x$. This factors out into x(x + 2).

You'll also need to know this about algebraic factoring:

- (x + y)(x + y) is the same as $x^2 + xy + xy + y^2$ because each item in parentheses gets multiplied separately by everything else in the other parentheses. This adds to $x^2 + 2xy + y^2$.
- (x + y)(x − y) becomes $x^2 + xy − xy + y^2$ or $x^2 − y^2$.
- (x − y)(x − y) becomes $x^2 − 2xy + y^2$.
- $x^3 + y^3$ becomes $(x + y)(x^2 - xy + y^2)$.
- $x^3 − y^3$ becomes $(x − y)(x^2 + xy + y^2)$.

It's all about finding common ways to find square roots to separate or "factor" the numbers and variables). We'll start with some relatively easy ones and work our way up to harder ones.

- **Problem 36.** Factor $2x^2 - 32 = 0$.
  Start by factoring out the 2 as a greatest common factor to make $2(x^2 - 16)$. You know that 16 is 4^2. Looking at the rules for algebraic factoring you get this: $2(x + 4)(x - 4) = 0$. Now, what do you do? This seems unsolvable. Except...think about it. If this is three things multiplied by each other and the answer is 0, one or both of the expressions in parentheses must equal to zero. This means that $x + 4 = 0$ or $x - 4 = 0$. But which is it? Actually, both are true and the real answer is $x = 4, -4$. This is how you describe the situation when x can be more than one term.

- **Problem 37.** Factor $3x^2 - 300 = 0$.
  Start by factoring the 3 to get $3(x^2 - 100) = 0$. Again, you have to know your square roots. The square root of 100 is 10 because 10^2 is 100. This leads again to $3(x^2 - 10^2) = 0$.
$$\text{Solve } 3(x^2 - 10^2) = 0.$$
$$3(x + 10)(x - 10) = 0$$
  Again, one of these factored terms must be zero for this equation to be true.
$$\text{So, } x + 10 = 0 \text{ or } x - 10 = 0$$
$$x = -10, 10 \text{ (either answer is correct)}.$$

- **Problem 38.** Factor $x^2 + 4x = -4$ and solve for x.
$$\text{Turn it into this: } x^2 + 4x + 4 = 0$$
  Following the rule of $x^2 + 2xy + y^2$ you get $(x + 2)(x + 2) = 0$
$$x \text{ must be -2}.$$
  To check this, put it in the equation: $(-2)^2 + 4(-2) = -4$
$$4 - 8 = -4 \text{ (This is correct)}.$$

- **Problem 39.** Factor this and solve for x. $3x^2 + 18x = -27$
$$\text{Set the equation equal to zero: } 3x^2 + 18x + 27 = 0$$
$$\text{Factor out the 3: } 3(x^2 + 6x + 9) = 0$$
$$\text{This becomes } 3(x + 3)(x + 3) = 0$$
  This means that $x + 3$ must equal to zero so $x = -3$.

- **Problem 40.** Factor this and solve for x. $x^2 + 10x + 25 = 0$.
  This should be relatively easy now. You know what the square root of 25 is because $5^2 = 25$.
  This leads you to this: $(x + 5)(x + 5) = 0$.
  Based on what you know about factoring and multiplication: $x = -5$

- **Problem 41.** Factor this and solve for x: $2x^2 + 28x + 98 = 0$
  Start by factoring out the 2 as every term divides by 2: $2(x^2 + 14x + 49) = 0$
  Because $7^2 = 49$, you can factor this out to be $2(x + 7)(x + 7) = 0$
  $x + 7$ must equal zero so $x = -7$.

- **Problem 42.** As this will get more complicated in coming chapters, you'll need to know how to expand factors from their components. Let's do a couple of these:
  Expand $2(x + 2)(x - 3) = 0$:
  Multiply everything by everything else, starting with the parentheses:
  $$(x + 2)(x - 3) = x^2 - 3x + 2x - 6$$
  $$2(x^2 - 3x + 2x - 6) = 0$$
  $$2(x^2 - 2x - 6) = 0$$
  $$2x^2 - 4x - 12 = 0$$

- **Problem 43.** Expand these factors: $(x + 1)(x + 2)(x + 3) = 0$
  Multiply two of the parentheses first: $(x + 1)(x^2 + 3x + 2x + 6) = 0$
  Now use modified distribution: $(x + 1)(x^2 + 5x + 6) = 0$
  $$x^3 + 5x^2 + 6x + x^2 + 5x + 6 = 0$$
  Next combine like terms: $x^3 + 6x^2 + 11x + 6 = 0$

- **Problem 44.** Just once, we'll expand and factor something complicated so you can see how it's done:
  Factor and solve this equation: $(x^4 + 14x^3 + 49x^2)/x^2 = 0$.
  How do you begin with this?
  You can remember that you can divide each of terms in the numerator by the denominator x^2. This leads to this: $x^4/x^2 + 14x^3/x^2 + 49x^2/x^2$ becomes this: $x^2 + 14x + 49$. You can further factor this:
  $$x^2 + 14x + 49 = 0$$
  $$(x + 7)(x + 7) = 0$$
  $$x = -7$$

These next questions in this chapter brings us back to word problems and primarily to those word problems that involve 2 - 3 variables that you don't know the answer to. You'll need to use algebra to create two to three separate equations that can be solved together. This leads to some of the same types of equations as we've just worked with, except that you'll use two or three variables to solve the problem. Let's give it a try...

- **Problem 45.** Person A and Person B made $250 in a raffle. Person A made x dollars and person B made y dollars. Person B bragged that he made $10 more than twice the number of dollars as person A. How many dollars did person A and person B make?
  $$\text{Equation 1: } x + y = \$250$$
  $$\text{Equation 2: } 2x + 10 = y$$
  The rest is pretty simple: $y = 2x + 10$
  Solve the first equation knowing this: $(2x + 10) + x = 250$
  $$3x + 10 = 250$$
  $$3x = 240$$
  $$x = 80$$
  This means that person A made $80 and person B made $170.

- **Problem 46.** You have a rectangle with an area of 36 square centimeters. Side x and side y are the two sides. You know that side y is 4 times side x. How do you determine what the length of the two sides are and what is the perimeter?

Equation 1: xy = 36 (the area of a rectangle)

Equation 2: is 4x = y

Equation 3: Perimeter = x + x = y + y

Solve the first equation, with what you know from the second:

$$x(4x) = 36$$
$$4x^2 = 36$$

Now divide by 4 to get: $x^2 = 9$

$$x = 3$$

Side x is 3 cm while side y is 12 cm

The perimeter of the rectangle is 30 cm.

- **Problem 47.** You have a rectangle with an area of 288 cm. The length of side x is twice the length of side y. What is the length of each side?

Equation 1: xy = 288

Equation 2: x = 2y

$$2y(y) = 288$$
$$2y^2 = 144$$
$$y = 12$$

The length of side y is 12 centimeters and the length of side x is 24 cm.

- **Problem 48.** Three people ran a course. Person A ran x miles, person B ran y miles, and person C ran z miles. Together they ran a total of 91 miles. Person B ran 2 miles farther than twice person A and 3 miles farther than person C. How many miles did each person run?

Equation 1: $y = 2x + 2$

Equation 2: $y - z = 3$

Equation 3: $x + y + z = 91$

You can start with any of these but since you know $y = 2x + 2$, put it into equation 2:

$(2x + 2) - z = 3$

$-z = -2x + 1$ (by moving the $-2x + 2$ to the righthand side) or $z = 2x - 1$

Equation 3: $x + (2x + 2) + (2x - 1) = 91$

$5x + 1 = 91$

$5x = 90$

$x = 18$ miles

This means that person A ran 18 miles, person B ran 38 miles, and person C ran 35 miles.

- **Problem 49.** The perimeter of a rectangle is 202 cm. Side A is 2 cm longer than twice side B. What is the length of side A and side B?

Equation 1: $2A + 2B = 202$

Equation 2: $A = 2 + 2B$

$2(2 + 2B) + 2B = 202$

$4 + 4B + 2B = 202$

$6B = 202 - 4$

$6B = 198$

$B = 33$

Side A is $2 + 2(33) = 68$ cm

Side B is 33 cm

- **Problem 50.** Person A raised x dollars for charity, while person B raised 4 dollars more than 3 times person B. Person B raised 5 dollars more than person C. Together, all three raised $1004 for charity. How much did each person raise?

Equation 1: $3x + 4 = y$

Equation 2: $y = z + 5$

Equation 3: $x + y + z = 1004$

Solve for z in equation 2: $z = y - 5$

Insert equation 1 into equation 2: $z = (3x + 4) - 5$

Remember, when there are three variables, there must be three equations to solve.

$x + (3x + 4) + [(3x + 4) -5] = 1004$

$x + 3x + 3x + 4 + 4 - 5 = 1004$

$7x = 1004 - 3$

$7x = 1001$

$x = 143$

Person A raised $143, Person B raised $433, and person C raised $428

CHAPTER 3:
THE DREADED QUADRATIC FORMULA

If you've heard of the quadratic formula or have seen it, you know it looks like a lot to swallow. It looks complicated but, in reality, it's a great tool for figuring out polynomial equations that otherwise can't easily be factored. The reason why we looked at factoring in the last chapter is because it sets the stage for this chapter. In the last chapter, we factored and expanded some easy equations—equations where there was a single number that could be factored out of the equation. But what if it isn't easy? What if there are two solutions for x that aren't the same and the equation doesn't easily factor? Well, it turns out that the quadratic formula can help with that.

The quadratic formula is this:

$$x = \frac{-b \pm \sqrt{b^2 - 4ac}}{2a}$$

This solves the equation ax^2 + bx + c = 0. Notice the plus or minus sign in the formula. This basically means that there will likely be two different solutions for x. It also uses the same formula you've been exposed to in the previous chapter, in which the quadratic equation has "0" on the right-hand side of the equation. This is because when you factor this out, such as (x -1)(x + 1) = 0, you must assume that either x + 1 = 0 or x – 1 = 0, which allows you to solve the equation. It means that, regardless of how the polynomial is laid out, you must transform it into an equation that looks like this:

$$ax^2 + bx + c = 0$$

You've already seen that there can be two values of x that work in these equations. This is where the plus-minus sign comes from in the quadratic formula. There will be two values of x in most situations. Let's get started and get through these equations!

- **Problem 51.** Solve this equation: $x^2 + 5x = 20$.

   Start by putting it into the form we can solve: $x^2 + 5x - 20 = 0$. This isn't easy to factor so you need to put this into the quadratic formula to solve it.

$$x = \frac{-5 \pm \sqrt{5^2 - 4(1)(-20)}}{2}$$

Remember that, for this equation, a = 1, b = 5, and c = -20. Separate this out:

$$x = \frac{-5+\sqrt{5^2-4(1)(-20)}}{2} \text{ and}$$

$$x = \frac{-5 - \sqrt{5^2 - 4(-20)}}{2}$$

Now just solve these equations separately:

$$x = \frac{-5+\sqrt{5^2-4(-20)}}{2} \text{ or } x = \frac{-5+\sqrt{25+80}}{2} \text{ or } x = \frac{-5+\sqrt{105}}{2} \text{ or x =}$$

$$\frac{-5}{2} + \frac{\sqrt{105}}{2}$$

And also: x = $\dfrac{-5}{2} - \dfrac{\sqrt{105}}{2}$

Now that was a little bit messy but it gives you the answer and, with a calculator, you can actually solve these two values of x. Now let's try something a little bit simpler.

- **Problem 52.** Solve this problem: $3x^2 + 42x = -147$

   You can try to factor this out but, as you've seen, there is always the quadratic formula to plug the values of a, b, and c into the equation that now looks like this: $3x^2 + 42x + 147 = 0$ with a =3, b = 42 and c = 147. To make it simple, let's factor this out to be $3(x^2 + 14x + 49) = 0$. Let's give it a try by doing the quadratic equation on just the values in the parentheses in which a = 1, b = 14, and c = 49:

$$x = \frac{-14 \pm \sqrt{14^2 - 4(1)(49)}}{2(1)}$$

$$x = \frac{-14+\sqrt{14^2-4(1)(49)}}{2(1)} \text{ and}$$

$$x = \frac{-14 - \sqrt{14^2 - 4(1)(49)}}{2(1)}$$

Now get out your calculator and figure this out:

$$x = \frac{-14 + \sqrt{196 - 196}}{2} \text{ and}$$

$$x = \frac{-14 - \sqrt{196 - 196}}{2}$$

$$x = -\frac{-14}{2} + \frac{\sqrt{0}}{2} \text{ and x} = \frac{-14}{2} - \frac{\sqrt{0}}{2}$$

Solving the problem, you get x = $\frac{-14}{2}$ or x = -7

- **Problem 53.** Solve this one: $x^2 + 3x - 4 = 0$
  In this problem a = 1, b = 3, and c = -4. Let's tackle the quadratic equation:

  $$x = \frac{-3 \pm \sqrt{3^2 - 4(-4)}}{2} \text{ or this:}$$

  $$x = \frac{-3 + \sqrt{3^2 - 4(-4)}}{2} \text{ and x} = \frac{-3 - \sqrt{3^2 - 4(-4)}}{2}$$

  You might not need your calculator for this one: x = $\frac{-3 + \sqrt{25}}{2}$ and x =

  $$\frac{-3 - \sqrt{25}}{2} \text{ or}$$

  $$x = \frac{-3 + 5}{2} \text{ and x} = \frac{-3 - 5}{2} \text{ or x = 1 and x = -4}$$

You've seen the simple and the not so simple with regard to the quadratic equation. The take home message is that it doesn't really matter what the equation is, as long as you put the right numbers into the quadratic formula, you'll get an answer.

- **Problem 54.** Solve $x^2 - 2x = 3$
  This translates to $x^2 - 2x - 3 = 0$; remember, we can use the quadratic formula, which only works for equations that are equal to zero.

  $$x = \frac{2 \pm \sqrt{(-2)^2 - 4(1)(-3)}}{2(1)} \text{ which becomes this:}$$

  $$x = \frac{2 + \sqrt{(-2)^2 - 4(1)(-3)}}{2(1)} \text{ and x} = \frac{2 - \sqrt{(-2)^2 - 4(1)(-3)}}{2(1)}$$

  $$x = \frac{2 + \sqrt{4 + 12}}{2} \text{ and x} = \frac{2 - \sqrt{4 + 12}}{2} \text{ or x} = \frac{2 + 4}{2} \text{ or x} = \frac{2 - 4}{2}$$

  Doing the math, you get x = 3, x = -1

- **Problem 55.** Solve $2x^2 + 16x = -32$

  This becomes $2x^2 + 16x + 32 = 0$. Now factor out 2 to get it in the format that looks like this $2(x^2 + 8x + 16) = 0$. Now do the quadratic formula, knowing a = 1, b = 8, and c = 16. Solve the quadratic formula like this:

  $$x = \frac{-8 \pm \sqrt{64 - 4(16)}}{2}, \text{ which leads to this:}$$

  $$x = \frac{-8 + \sqrt{64 - 4(16)}}{2} \text{ and } x = \frac{-8 - \sqrt{64 - 4(16)}}{2}$$

  Under the square root, you get $64 - 64 = 0$

  $$x = \frac{-8}{2} \text{ and } x = \frac{-8}{2} \text{ or x = -4}$$

- **Problem 56.** Solve $x^2 - 5x + 6 = 0$

  This should be easy now: a = 1, b = -5, and c = 6.

  $$x = \frac{5 \pm \sqrt{(-5)^2 - 4(6)}}{2} \text{ which leads to this:}$$

  $$x = \frac{5 + \sqrt{(-5)^2 - 4(6)}}{2} \text{ and } x = \frac{5 - \sqrt{(-5)^2 - 4(6)}}{2}$$

  $$x = \frac{5 + 1}{2} \text{ and } x = \frac{5 - 1}{2}$$

  This leads to x = 3 and x = 2

- **Problem 57.** Solve $x^2 - 6x = -8$

  In this equation, $x^2 - 6x + 8 = 0$ and a = 1, b = -6, and c = 8.

  $$x = \frac{6 \pm \sqrt{(-6)^2 - 4(8)}}{2} \text{ which leads to this:}$$

  $$x = \frac{6 + \sqrt{(-6)^2 - 4(8)}}{2} \text{ and } x = \frac{6 - \sqrt{(-6)^2 - 4(8)}}{2}$$

  This leads to this:

  $$x = \frac{6 + \sqrt{36 - 32)}}{2} \text{ and } x = \frac{6 - \sqrt{36 - 32}}{2}$$

  This leads to this:

  $$x = \frac{6 + \sqrt{4}}{2} \text{ and } x = \frac{6 - \sqrt{4}}{2}$$

  $$x = \frac{8}{2} \text{ and } x = \frac{4}{2}$$

  x = 4 and x = 2

- **Problem 58.** Solve $5x^2 - 20x = -20$

  Factor out 5 to get $5(x^2 -4x + 4) = 0$, to get a = 1, b = -4, and c = 4. After this, you can ignore the 5 because, if you divide both sides of the equation by 5, you still get "0" on the right side of the equation.

$$x = \frac{-(-4)\pm\sqrt{(-4)^2-4(4)}}{2}$$ or this:

$$x = \frac{-(-4)+\sqrt{(-4)^2-4(4)}}{2} \text{ and } = \frac{-(-4)-\sqrt{(-4)^2-4(4)}}{2} \text{ , which leads to:}$$

$$x = \frac{4+\sqrt{0}}{2} \text{ and } x = \frac{4-\sqrt{0}}{2}$$

$$x = 2$$

- **Problem 59.** Solve $x^2 -7x +12 = 0$

  In this equation, a = 1, b = -7, and c = 12.

$$x = \frac{-(-7)\pm\sqrt{(-7)^2-4(12)}}{2}$$ which leads to this:

$$x = \frac{7+\sqrt{49-48}}{2} \text{ and } x = \frac{7-\sqrt{49-48}}{2}$$

$$x = \frac{7+1}{2} \text{ and } x = \frac{7-1}{2}$$

The answers are then that x = 4 and x = 3

- **Problem 60.** Solve $(x -4)(x + 2) = 32$

  Expand this factored quadratic to get $x^2 - 2x - 8 - 32 = 0$ or $x^2 - 2x - 40 = 0$. In this equation, you get a = 1, b = -2, and c = -40.

$$x = \frac{-(-2)\pm\sqrt{(-2)^2-4(-40)}}{2}$$ which leads to this:

$$x = \frac{-(-2)+\sqrt{(-2)^2-4(-40)}}{2} \text{ and } x = \frac{-(-2)-\sqrt{(-2)^2-4(-40)}}{2}$$

$$x = \frac{2+\sqrt{4+160}}{2} \text{ and } x = \frac{2-\sqrt{4+160}}{2}$$

This is a little bit more complicated but in can be done:

$$\frac{2+\sqrt{164}}{2} \text{ and } \frac{2-\sqrt{164}}{2} \quad \frac{2+2\sqrt{41}}{2} \text{ and } \frac{2-2\sqrt{41}}{2}$$

This simplifies into $x = 1 + \sqrt{41}$ and $x = 1 - \sqrt{41}$ because you can divide by 2.

- **Problem 61.** Solve $x^2 + 10x + 25 = 0$

  In this equation, you get a = 1, b = 10, and c = 25.

  $$x = \frac{-10 \pm \sqrt{10^2 - 4(25)}}{2}$$ or this:

  $$x = \frac{-10 + \sqrt{0}}{2} \text{ and } x = \frac{-10 - \sqrt{0}}{2}$$

  $$x = -5$$

- **Problem 62.** Solve $x^2 + x = 6$

  Change the equation to $x^2 + x - 6 = 0$. In this equation, a = 1, b = 1, and c = -6.

  $$x = \frac{-1 \pm \sqrt{1^2 - 4(-6)}}{2}$$ which leads to this:

  $$x = \frac{-1 + \sqrt{1^2 - 4(-6)}}{2} \text{ and } x = \frac{-1 - \sqrt{1^2 - 4(-6)}}{2}$$

  $$x = \frac{-1 + \sqrt{1^2 + 24}}{2} \text{ and } x = \frac{-1 - \sqrt{1^2 + 24}}{2}$$

  $$x = \frac{-1 + \sqrt{25}}{2} \text{ and } x = \frac{-1 - \sqrt{25}}{2}$$

  $$x = \frac{-1 + 5}{2} \text{ and } x = \frac{-1 - 5}{2}$$

  $$x = \frac{4}{2} \text{ and } x = \frac{-6}{2}$$

  $$x = 2 \text{ and } x = -3$$

- **Problem 63.** Solve $x^2 - 5x = 14$

  Change the equation to $x^2 - 5x - 14 = 0$. In this equation a = 1, b = -5, and c = -14.

  $$x = \frac{5 \pm \sqrt{(-5)^2 - 4(-14)}}{2}$$ which becomes this:

  $$x = \frac{5 + \sqrt{(-5)^2 - 4(-14)}}{2} \text{ and } x = \frac{5 - \sqrt{(-5)^2 - 4(-14)}}{2}$$

  $$x = \frac{5 + \sqrt{25 + 56}}{2} \text{ and } x = \frac{5 - \sqrt{25 + 56}}{2}$$

  $$x = \frac{5 + \sqrt{81}}{2} \text{ and } x = \frac{5 - \sqrt{81}}{2}$$

  $$x = \frac{5 + 9}{2} \text{ and } x = \frac{5 - 9}{2}$$

  $$x = 7 \text{ and } x = -2$$

- **Problem 64.** Solve $x^2 + 11x + 18 = 0$

  In this equation, you get a = 1, b = 11, and c = 18.

  $$x = \frac{-11 \pm \sqrt{11^2 - 4(18)}}{2}$$ so, get out your calculator to get this:

  $$x = \frac{-11 + \sqrt{121 - 72}}{2} \text{ and } x = \frac{-11 - \sqrt{121 - 72}}{2}$$

  $$x = \frac{-11 + \sqrt{49}}{2} \text{ and } x = \frac{-11 - \sqrt{49}}{2}$$

  $$x = \frac{-4}{2} \text{ and } x = \frac{-18}{2}$$

  x = -2 and x = -9

- **Problem 65.** Solve $x^2 - 11x = -30$

  Change the equation to $x^2 - 11x + 30 = 0$. In this equation, a = 1, b = -11, and c = 30.

  $$x = \frac{11 \pm \sqrt{(-11)^2 - 4(30)}}{2}$$ and continuing from there, you get:

  $$x = \frac{11 + \sqrt{121 - 120)}}{2} \text{ and } x = \frac{11 - \sqrt{121 - 120}}{2}$$

  $$x = \frac{11 + 1}{2} \text{ and } x = \frac{11 - 1}{2}$$

  x = 6 and x = 5

- **Problem 66.** Solve $3x^2 - 9x = 30$

  Change the equation to $3x^2 - 9x - 30 = 0$. In this equation, you can divide by 3 to get $3(x^2 - 3x - 10) = 0$. It leaves a = 1, b = -3, and c = -10.

  $$x = \frac{3 \pm \sqrt{(-3)^2 - 4(-10)}}{2}$$ which leads to this:

  $$x = \frac{3 + \sqrt{9 + 40}}{2} \text{ and } x = \frac{3 - \sqrt{9 + 40}}{2}$$

  $$x = \frac{3 + 7}{2} \text{ and } x = \frac{3 - 7}{2}$$

  $$x = \frac{10}{2} \text{ and } x = \frac{-4}{2}$$

  x = 5 and x = -2

- **Problem 67.** Solve $x^2 - 20x = -100$

  Change the equation to $x^2 - 20x + 100 = 0$. In this equation, a = 1, b = -20, and c = 100.

$$x = \frac{20 \pm \sqrt{(-20)^2 - 4(100)}}{2}$$ which becomes this:

$$x = \frac{20 + \sqrt{400 - 400}}{2} \text{ and } x = \frac{20 - \sqrt{400 - 400}}{2}$$

$$x = \frac{20 + \sqrt{0}}{2}$$

x = 10

- **Problem 68.** Solve $x^2 + 3x = 10$

  Change the equation to $x^2 + 3x - 10 = 0$. In this equation, a = 1, b = 3, and c = -10.

$$x = \frac{-3 \pm \sqrt{3^2 - 4(-10)}}{2}$$ which easily becomes this:

$$x = \frac{-3 + \sqrt{49}}{2} \text{ and } x = \frac{-3 - \sqrt{49}}{2}$$

$$x = \frac{-3 + 7}{2} \text{ and } x = \frac{-3 - 7}{2}$$

$$x = \frac{4}{2} \text{ and } x = \frac{-10}{2}$$

x = 2 and x = -5

- **Problem 69.** Solve $2x^2 + 12x = 32$

  Change the equation to $2x^2 + 12x - 32 = 0$. Divide by 2 to get $2(x^2 + 6x - 16) = 0$, which leads to a = 1, b = 6 and c = -16

$$x = \frac{-6 \pm \sqrt{6^2 - 4(-16)}}{2}$$ and this leads to this:

$$x = \frac{-6 + \sqrt{36 + 64}}{2} \text{ and } x = \frac{-6 - \sqrt{36 + 64}}{2}$$

$$x = \frac{-6 + \sqrt{100}}{2} \text{ and } x = \frac{-6 - \sqrt{100}}{2}$$

$$x = \frac{-6 + 10}{2} \text{ and } x = \frac{-6 - 10}{2}$$

x = 2 and x = -8

- **Problem 70.** Solve $x^2 - 9x + 18 = 0$
  In this equation, a = 1, b = -9, and c = 18.

$$x = \frac{9 \pm \sqrt{(-9)^2 - 4(18)}}{2}$$ which, by now you know means this:

$$x = \frac{9 + \sqrt{81-72}}{2} \text{ and } x = \frac{9 - \sqrt{81-72}}{2}$$

$$x = \frac{9 + \sqrt{9}}{2} \text{ and } x = \frac{9 - \sqrt{9}}{2}$$

$$x = \frac{9+3}{2} \text{ and } x = \frac{9-3}{2}$$

x = 6 and x = 3

Now we'll do a few word problems involving polynomial equations using the quadratic formula. The quadratic formula works, not just for simple equations. You can do just about any word problem with 2 variables and 2 equations using simple math and the quadratic formula.

- **Problem 71.** Solve this using algebra. You have a rectangle with an area of 200 cm^2 and you know that one side is 6 inches longer than the other side. To the nearest hundredth, what are the lengths of the sides?

Equation is: x(x + 6) = 200

$$x^2 + 6x - 200 = 0$$

a = 1, b = 6, c = -200

$$x = \frac{-6 \pm \sqrt{6^2 - 4(-200)}}{2}$$ which translates to this:

$$x = \frac{-6 + \sqrt{36+800}}{2} \text{ and } x = \frac{-6 - \sqrt{36+800}}{2}$$

$$x = \frac{-6 + 28.9}{2} \text{ and } x = \frac{-6 - 28.9}{2}$$

The second x cannot be the right answer because a length cannot be a negative number. This means the first one is the correct answer: x = 22.9/2 or x = 11.45 cm and y = 17.45 cm.

- **Problem 72.** Solve this using algebra. You have a rectangle with an area of 400 cm^2 and you know that side y is 4 inches longer than twice the length of x. To the nearest tenth, what are the lengths of the sides?

The equation: $x(2x + 4) = 400$

$$2x^2 + 4x - 400 = 0$$

$$a = 2, b = 4, c = -400$$

$x = \frac{-4 \pm \sqrt{4^2 - 4(2)(-400)}}{2(2)}$ which becomes this:

$$x = \frac{-4 + \sqrt{16 + 3200}}{2(2)} \text{ and } x = \frac{-4 - \sqrt{16 + 3200}}{2(2)}$$

$$x = \frac{-4 + \sqrt{3216}}{4} \text{ and } x = \frac{-4 - \sqrt{3216}}{4}$$

$$x = \frac{-4 + 56.7}{4} \text{ and } x = \frac{-4 - 56.7}{4}$$

According to the rules, this cannot be a negative number so:

$$x = \frac{52.7}{4}$$

x = 13.2 cm and y = 2(13.2) + 4 or y = 30.4 cm

- **Problem 73.** Solve this using algebra. You have a rectangle that has an area of 500 cm^2 and has one side that is 2 inches longer than 3 times the other side. To the nearest hundredths, what are the lengths of the sides?

The equation is $(3x + 2)x = 500$

$$3x^2 + 2x - 500 = 0$$

This means a = 3, b = 2, and c = -500

$x = \frac{-2 \pm \sqrt{(-2)^2 - 4(3)(-500)}}{2(3)}$ and following through, you get this:

$$x = \frac{-2 + \sqrt{4 + 6000}}{2(3)} \text{ and } x = \frac{-2 - \sqrt{4 + 6000}}{2(3)}$$

$x = \frac{-2 + 77.5}{6}$ and $x = \frac{-2 - 77.5}{6}$ (but it can't be negative) so:

$x = \frac{-2 + 77.5}{6} = 75.5/6$ or x = 12.58 cm and y = (3)(12.58) + 2 or y = 39.74 cm

- **Problem 74.** Solve this using algebra. George has a pool with an area of 200 square feet. Side x is shorter and side y is 5 feet longer than twice side x. What are the dimensions of the pool?

$$\text{Equation: } x(2x + 5) = 200$$
$$2x^2 + 5x - 200 = 0$$

This means that a = 2, b = 5, and c = -200

$$x = \frac{-5 \pm \sqrt{5^2 - 4(2)(-200)}}{2(2)} \text{ which becomes this:}$$

$$x = \frac{-5 + \sqrt{25 + 1600}}{2(2)} \text{ and } x = \frac{-5 - \sqrt{25 + 1600}}{2(2)}$$

$$x = \frac{-5 + 40.3}{4} \text{ and } x = \frac{-5 - 40.3}{4} \text{ but since it can't be negative:}$$

$$x = \frac{35.3}{4}$$

x = 8.8 feet and y = 22.6 feet

- **Problem 75.** Solve this using algebra. Mary has a garden that is rectangular. Side y is longer than side x. In fact, it is 5 feet longer than side x. The area of the garden is 75 square feet. What is the perimeter of the garden?

$$\text{The equation is this: } x(x + 5) = 75$$
$$x^2 + 5x - 75 = 0$$

So, this means that a = 1, b = 5, and c = -75

$$x = \frac{-5 \pm \sqrt{5^2 - 4(-75)}}{2} \text{ which leads to this:}$$

$$x = \frac{-5 + \sqrt{25 + 300}}{2} \text{ and } x = \frac{-5 - \sqrt{25 + 300}}{2} \text{ but, as it can't be negative:}$$

$$x = \frac{-5 + \sqrt{325}}{2} \text{ or x = 6.5 feet and y = 11.5 feet}$$

The perimeter is 6.5 + 11.5 + 6.5 + 11.5 = 36 feet

CHAPTER 4:
RATIONAL EQUATIONS AND BIQUADRATIC EQUATIONS

Rational equations are equations containing at least one fraction whose numerator and denominator are polynomials. There is a special way to do this that will make these types of equations very simple to do. Look at something simple like this: $\frac{1}{x} = 2$. This is a simple fraction equation that can also be written like this: $\frac{1}{x} = \frac{2}{1}$. The next thing you need to know is that in any fraction equation, such as $\frac{a}{b} = \frac{c}{d}$, this is also true: ad = bc. In the case of the equation in question, then 1 = 2x (just cross-multiply the fractions and you get this equation). Now, you can say that x = ½. We'll use this fraction situation and the rules of fraction solving as it applies to polynomial fractions and rational equations.

Biquadratic equations will be covered as well in this chapter. These are equations that have polynomials of a higher order than simply x^2. While solving the equation like $x^4 - 5x^2 + 4 = 0$ seems extra complicated, these equations may come up as you study algebra so we should cover them. They really aren't as difficult as they initially seem. These types of questions are totally solvable. Let's get started!

- **Problem 76.** Solve this equation. $\frac{x+2}{x-2} = \frac{x+4}{x+1}$

This multiplies into this: $(x + 2)(x +1) = (x - 2)(x + 4)$
$$x^2 + x + 2 + 2x = x^2 + 4x - 2x - 8$$
$$x^2 + x + 2 + 2x - x^2 - 4x + 2x + 8 = 0$$
$$x + 10 = 0$$
$$x = -10$$

- **Problem 77.** Solve this equation. $\dfrac{x-1}{2x-2} = \dfrac{2x-1}{x+2}$

  This multiplies out into this equation: $(2x - 2)(2x - 1) = (x -1)(x + 2)$

  Expand the equation: $4x^2 - 2x - 4x + 2 = x^2 - x + 2x - 2$

  $4x^2 - 2x - 4x + 2 - x^2 + x - 2x + 2 = 0$

  $3x^2 - 7x + 4 = 0$

  It's time to break out the quadratic formula with a = 3, b = -7, c = 4

  $x = \dfrac{7 \pm \sqrt{(-7)^2 - 4(3)(4)}}{2(3)}$ which simplifies to this:

  $x = \dfrac{7 + \sqrt{49 - 48}}{2(3)}$ and $x = \dfrac{7 - \sqrt{49 - 48}}{2(3)}$

  $x = \dfrac{7 + \sqrt{1}}{2(3)}$ and $x = \dfrac{7 - \sqrt{1}}{2(3)}$

  x = -8/6 and x = 6/6

  x = 4/3 and x = 1

  However, if $x = 1$, the fraction on the left is undefined because you cannot divide by zero, so the only answer is

  $$x = \frac{4}{3}$$

- **Problem 78.** Solve this equation. $\dfrac{2x-3}{2x+2} = \dfrac{x+2}{x-1}$

  Multiply it out: $(2x - 3)(x - 1) = (2x + 2)(x + 2)$

  Expand it: $2x^2 - 3x - 2x + 3 = 2x^2 + 4x + 2x + 4$

  Simplify it: $2x^2 - 3x - 2x + 3 - 2x^2 - 4x - 2x - 4 = 0$

  -11x - 1 = 0

  -11x = 1

  x = -1/11

- **Problem 79.** Solve this equation. $\frac{2x-4}{x-1} - \frac{3x-4}{x-2}$

Multiply it out: $(2x - 4)(x - 2) = (x - 1)(3x - 4)$

Expand it: $2x^2 - 2x - 4x + 8 = 3x^2 - 4x - 3x + 4$

Simplify it: $2x^2 - 2x - 4x + 8 - 3x^2 + 4x + 3x - 4 = 0$

$-x^2 + x + 4 = 0$

Put in the quadratic equation with a = -1, b = 1, c = 4

$x = \frac{-1 \pm \sqrt{1^2 - 4(-1)(4)}}{2(-1)}$ which translates into this:

$x = \frac{-1 + \sqrt{1+16}}{-2}$ and $x = \frac{-1 - \sqrt{1+16}}{-2}$ This is starting to look messy but it's solvable.

x = ½ - √17/2 and x = ½ + √17/2

You can get out your calculator if this is too messy to get this:

x = 0.5 − 4.1/2 and x = 0.5 + 4.1/2

x = -1.56 and x = 2.56

- **Problem 80.** Solve this equation. $\frac{5}{x-7} = \frac{2}{x-2}$

Multiply it out: $5(x - 2) = (x - 7)2$

Expand it: $5x - 10 = 2x - 14$

Simplify it: $3x + 4 = 0$

$3x = -4$

$x = -4/3$

- **Problem 81.** Let's try something that looks hard but really isn't. Solve this equation.

$$\frac{7x+5}{x-4} - \frac{2x+3}{x-1} = 0$$

Simplify it: $\frac{7x+5}{x-4} = \frac{2x+3}{x-1}$

Multiply it out: $(7x + 5)(x -1) = (x -4)(2x + 3)$

Expand it: $7x^2 -7x + 5x - 5 = 2x^2 + 3x - 8x - 12$

$$7x^2 -7x + 5x - 5 - 2x^2 - 3x + 8x + 12 = 0$$

$$5x^2 + 3x +7 = 0$$

Use the quadratic formula: a = 5, b =3, and c = 12

$$x = \frac{-3 \pm \sqrt{(3)^2 - 4(5)(12)}}{2(5)}$$ which translates to this:

$$x = \frac{3 + \sqrt{9-240}}{10} \text{ and } x = \frac{3 - \sqrt{9-240}}{10}$$

Simplify: $x = \frac{3 + \sqrt{-231}}{10}$ and $x = \frac{3 - \sqrt{-241}}{10}$

Notice the radical has a negative number inside. Therefore, there is no real solution to the given equation.

- **Problem 82.** Solve this equation. $\frac{2x-1}{4x-2} - \frac{x-2}{2x-1} = 0$

$$\frac{2x-1}{4x-2} = \frac{x-2}{2x-1}$$

Multiply it out: $(2x - 1)(2x -1) = (4x - 2)(x - 2)$

Expand it: $4x^2 - 2x - 2x + 1 = 4x^2 - 8x - 2x + 4$

$$4x^2 - 2x - 2x + 1 - 4x^2 + 8x + 2x - 4 = 0$$

Simplify it: $6x - 3 = 0$

$$6x = 3$$

x = 3/6 or x = ½

However, if $x = 1/2$, the denominators of both fractions will equal zero, which means the given equation has no solutions.

- **Problem 83.** Solve this equation. $\frac{8x+2}{x+4} = \frac{4x-2}{x-1}$

    Multiply it out: $(8x + 2)(x - 1) = (x + 4)(4x - 2)$

    Simplify it: $8x^2 - 8x + 2x - 2 = 4x^2 - 2x + 16x - 8$

    $8x^2 - 8x + 2x - 2 - 4x^2 + 2x - 16x + 8 = 0$

    $4x^2 - 20x + 6 = 0$

    Divide the whole thing by 2 to get: $2x^2 - 10x + 3 = 0$

    Use the quadratic formula with a = 2, b = -10, c = 3

    $$x = \frac{10 \pm \sqrt{(-10)^2 - 4(2)(3)}}{2(2)}$$ which leads to this:

    $$x = \frac{10 + \sqrt{100 - 24}}{4} \text{ and } x = \frac{10 - \sqrt{100 - 24}}{4}$$

    $$x = \frac{10 + \sqrt{76}}{4} \text{ and } x = \frac{10 - \sqrt{76}}{4}$$

    $$x = \frac{10 + 2\sqrt{19}}{4} \text{ and } x = \frac{10 - 2\sqrt{19}}{4}$$

    $$x = \frac{5}{2} + \frac{\sqrt{19}}{2} \text{ and } \frac{5}{2} - \frac{\sqrt{19}}{2}$$

Okay, so it's messy but it's nevertheless the answer that fits. It just goes to show that basically anything is solvable using these methods, even if the answer is messy.

- **Problem 84.** Solve this equation. $\frac{7}{x-3} - \frac{10}{x-2} - \frac{6}{x-1} = 0$ Okay, this is a little more complex but can be solved by finding the common denominator of the fractions:

The common denominator of each of these fractions is long but is (x -3)(x-2)(x -1).

Remember that you can multiply each of these fractions by the common denominator to get this:

$\frac{7(x-3)(x-2)(x-1)}{(x-3)(x-3)(x-2)(x-1)} - \frac{10(x-3)(x-2)(x-1)}{(x-2)(x-3)(x-2)(x-1)} -$
$\frac{6(x-3)(x-2)(x-1)}{(x-1)(x-3)(x-2)(x-1)} = 0$ Simplify it:

$$\frac{7(x-2)(x-1)}{(x-3)(x-2)(x-1)} - \frac{10(x-3)(x-1)}{(x-3)(x-2)(x-1)}$$
$$- \frac{6(x-3)(x-2)}{(x-3)(x-2)(x-1)} = 0$$

Now, since every term has the same denominator and the equation equals zero, the denominators can be cancelled, which gives this equation.

Simplify it further: $7(x - 2)(x - 1) - 10(x - 3)(x - 1) - 6(x -3)(x - 2) = 0$

Multiply it out: $7(x^2 - 2x - x + 2) - 10(x^2 - x - 3x + 3) - 6(x^2 - 2x - 3x + 6) = 0$

combine like terms to get: $7(x^2 - 3x + 2) - 10(x^2 - 4x + 3) - 6(x^2 - 5x + 6) = 0$

Distribute: $7x^2 - 21x + 14 - 10x^2 + 40x - 30 - 6x^2 + 30x - 36 = 0$

Combine like terms again: $-9x^2 + 49x - 52 = 0$

Let's solve this: a = -9, b = 49, c = -52

The dreaded quadratic formula gives this:

$x = \frac{-49 \pm \sqrt{49^2 - 4(-9)(-52)}}{2(-9)}$ which becomes this:

$x = \frac{-49 + \sqrt{2401 - 1872}}{-18}$ and $x = \frac{-49 - \sqrt{2401 - 1872}}{-18}$

$x = \frac{-49 + \sqrt{529}}{-18}$ and $x = \frac{-49 - \sqrt{529}}{-18}$

$x = \frac{-49 + 23}{-18}$ and $x = \frac{-49 - 23}{-18}$

$x = \frac{-26}{-18}$ and $x = \frac{-72}{-18}$

x = 13/9 and x = 4

Whew! That was a very long and complex problem but, if you do your simple math correctly and get your pluses and minuses straight, this is completely doable.

- **Problem 85.** Solve this equation. $\frac{x-3}{x-4} = \frac{x-6}{3x-1}$

    Multiply it out: $(x - 3)(3x - 1) = (x - 4)(x - 6)$

    Expand it: $3x^2 - x - 9x + 3 = x^2 - 6x - 4x + 24$

    Simplify it: $3x^2 - x - 9x + 3 - x^2 + 6x + 4x - 24 = 0$

    $$2x^2 - 21 = 0$$

    $$x^2 = \frac{21}{2}$$

    $$x = \pm \frac{\sqrt{21}}{\sqrt{2}}$$

    Now rationalize to remove the radical in the denominator.

    $$x = \pm \frac{\sqrt{21}}{\sqrt{2}} \cdot \frac{\sqrt{2}}{\sqrt{2}}$$

    $$x = \pm \frac{\sqrt{42}}{2}$$

    Again, this is messy but completely solvable.

Now we'll do some biquadratic equations, which have an extra step or two but can actually be done just as easily as any quadratic equation.

- **Problem 86.** Solve this equation. $x^4 - 5x^2 + 4 = 0$
First you have to create a placeholder variable in order to simplify the equation: the placeholder is z which we'll say $z = x^2$. This basically means that when you square both sides, you can get this: $z^2 = x^4$. Hopefully, you can see the usefulness to this and make this equation to solve:

$$z^2 - 5z + 4 = 0$$

Do the quadratic formula on this: a = 1, b = -5, c = 4

$$z = \frac{5 \pm \sqrt{(-5)^2 - 4(4)}}{2}$$ which leads to this:

$$z = \frac{5 + \sqrt{25 - 16}}{2} \text{ and } z = \frac{5 - \sqrt{25 - 16}}{2}$$

$$z = \frac{5 + \sqrt{9}}{2} \text{ and } z = \frac{5 - \sqrt{9}}{2}$$

$$z = \frac{5 + 3}{2} \text{ and } z = \frac{5 - 3}{2}$$

$$z = 4 \text{ and } z = 1$$

Since $z = x^2$ this leads to $x^2 = 4$ and $x^2 = 1$

Because $(-2)^2 = 4$, $2^2 = 4$, $1^2 = 1$, and $(-1)^2 = 1$, you have to include all four numbers in the answer.

$$x = 2, -2 \text{ and } x = 1, -1$$

- **Problem 87.** Solve this equation. $x^4 - 13x^2 + 36 = 0$
Start with $z = x^2$ and do this: $z^2 - 13z + 36 = 0$
Use the quadratic formula with a =1, b =-13, c =36

$$z = \frac{13 \pm \sqrt{(-13)^2 - 4(36)}}{2}$$ and then do this:

$$z = \frac{13 + \sqrt{169 - 144}}{2} \text{ and } z = \frac{13 - \sqrt{169 - 144}}{2}$$

$$z = \frac{13 + \sqrt{25}}{2} \text{ and } z = \frac{13 - \sqrt{25}}{2}$$

$$z = \frac{13 + 5}{2} \text{ and } z = \frac{13 - 5}{2}$$

$$z = \frac{13 + 5}{2} \text{ and } z = \frac{13 - 5}{2}$$

$$z = 9 \text{ and } z = 4$$

Since $z = x^2$ you take the square root of 9 and 4 to get this:

$$x = 3, -3 \text{ and } x = 2, -2$$

- **Problem 88.** Solve this equation. $4x^4 - 13x^2 + 9 = 0$
  Using the equation $z = x^2$ you get $4z^2 - 13z + 9 = 0$ that can be solved with the quadratic formula like this: $a = 4$ $b = -13$, $c = 9$

  $$z = \frac{13 \pm \sqrt{(-13)^2 - 4(4)(9)}}{2(4)}$$ which breaks down into this:

  $$z = \frac{13 + \sqrt{169 - 144}}{8} \text{ and } z = \frac{13 - \sqrt{169 - 144}}{8}$$

  $$z = \frac{13 + \sqrt{25}}{8} \text{ and } z = \frac{13 - \sqrt{25}}{8}$$

  $$z = \frac{13 + 5}{8} \text{ and } z = \frac{13 - 5}{8}$$

  $$z = \frac{9}{4} \text{ and } z = \frac{8}{8} = 1$$

  This means that $x = \sqrt{\frac{9}{4}} = 3/2$

  $x = 3/2, -3/2$ and $x = 1, -1$

- **Problem 89.** Find all solutions to this equation. $4x^4 + 3x^2 - 1 = 0$
  Using $z = x^2$ you get $4z^2 + 3z - 1 = 0$
  Use the quadratic formula to get this with a = 4, b = 3, and c = -1

  $$z = \frac{-3 \pm \sqrt{3^2 - 4(4)(-1)}}{2(4)}$$ which goes on to make this:

  $$z = \frac{-3 + \sqrt{9 + 16}}{8} \text{ and } z = \frac{-3 - \sqrt{9 + 16}}{8}$$

  $$z = \frac{-3 + \sqrt{25}}{8} \text{ and } z = \frac{-3 - \sqrt{25}}{8}$$

  $$z = \frac{-3 + 5}{8} \text{ and } z = \frac{-3 - 5}{8}$$

  $$z = \frac{2}{8} \text{ and } z = \frac{-8}{8}$$

  $$z = \frac{1}{4} \text{ and } z = -1$$

  This where we get into imaginary numbers. An imaginary number is what we get when we try to divide a negative number, such as -1. The square root of -1 is an imaginary number called *i* or *-i*. It doesn't make any real sense but this is what mathematicians do when they cannot solve a negative square root. So, the answer to the problem is this:

  Real numbers: x =1/2, -1/2 and imaginary numbers: x = *i* and *-i*

- **Problem 90.** Find all solutions to this equation. $x^4 + 25x^2 = 0$

  Okay, this one is deceptively simple (or so it seems). This is how it works: pretend that $z = x^2$ and that $z^2 = x^4$ plug it in to make this: $z^2 + 25z = 0$ or $z^2 = -25z$. One answer you probably haven't thought of is that z (and x) can be 0 because if you put 0 on either side of the equation, you still get the right answer (0 = 0). So, this means that one answer is x = 0. There are two others. Do this:

  $$\frac{z^2}{z} = \frac{-25}{z} \text{ which means } z = -25$$

  So, what does this mean for x? x is the square root of z so this means that $x = \sqrt{-25}$ or this:

  $X = \sqrt{25} \ (\sqrt{-1})$ This means that, using imaginary numbers again, x = 0, x = -5i and x = 5i

- **Problem 91.** Solve this equation. $(x^2 + 1)(x^2 - 5) = -8$

  Expand the equation: $x^4 - 5x^2 + x^2 - 5 + 8 = 0$

  $$x^4 - 4x^2 + 3 = 0$$

  Replace $z = x^2$ to get this: $z^2 - 4z + 3 = 0$

  Using the quadratic formula, you have a = 1, b = -4, c = 3

  $$z = \frac{-(-4) \pm \sqrt{(-4)^2 - 4(1)(3)}}{2(1)} \text{ which leads to this:}$$

  $$z = \frac{4 + \sqrt{16 - 12}}{2} \text{ and } z = \frac{4 - \sqrt{16 - 12}}{2}$$

  $$z = \frac{4 + \sqrt{4}}{2} \text{ and } z = \frac{4 - \sqrt{4}}{2}$$

  $z = (4+2)/2$ and $z = (4-2)/2$

  $z = 6/2$ and $z = \frac{2}{2}$ or z = 3 and z = 1

  Remembering the square root of z is x, you get this:

  $x = -\sqrt{3}, \sqrt{3}, 1, -1$

- **Problem 92.** Find all solutions to this equation: $x^4 - 4x^2 = 140$

  Substituting z for x^2 and z^2 for x^4, you get: $z^2 - 4z - 140 = 0$

  This leaves a = 1, b = -4, and c = -140:

  $$z = \frac{4 \pm \sqrt{(-4)^2 - 4(-140)}}{2}$$ which leads to this:

  $$z = \frac{4 + \sqrt{16 + 560}}{2} \text{ and } z = \frac{4 - \sqrt{16 + 560}}{2}$$

  $$z = \frac{4 + \sqrt{576}}{2} \text{ and } z = \frac{4 - \sqrt{576}}{2}$$

  $$z = \frac{4 + 24}{2} \text{ and } z = \frac{4 - 24}{2}$$

  $$z = 14 \text{ and } z = -10$$

  This gets into the imaginary number thing again:

  $$x = -\sqrt{14},\ \sqrt{14},\ -i\sqrt{10},\ i\sqrt{10}$$

- **Problem 93.** Solve this equation. $2x^4 - 10x^2 = 72$

  Using the substitution that we've been using, you get this: $2z^2 - 10x - 72 = 0$

  This leaves us with: a = 2, b = -10, c = -72

  Using the quadratic equation:

  $$z = \frac{10 \pm \sqrt{(-10)^2 - 4(2)(-72)}}{2(4)}$$ which leads to this:

  $$z = \frac{10 + \sqrt{100 + 576}}{8} \text{ and } z = \frac{10 - \sqrt{100 + 576}}{8}$$

  $$z = \frac{10 + \sqrt{676}}{8} \text{ and } z = \frac{10 - \sqrt{676}}{8}$$

  $$z = \frac{10 + 26}{8} \text{ and } z = \frac{10 - 26}{8}$$

  $$z = \frac{36}{8} \text{ and } z = \frac{-16}{8} \text{ or } z = 4 \text{ and } z = -2$$

  Again, it gets messy with imaginary numbers:

  $$x =, -2, 2, i\sqrt{2}, i\sqrt{2}$$

- **Problem 94.** Solve this equation. $x^3 - 2x = 0$ (This one looks hard but you can solve it!)

  Factor out the left side to get $x(x^2 - 2) = 0$. This means that x can be 0. It also means:

  $$x^2 - 2 = 0$$
  $$x = +/-\sqrt{2}$$

- **Problem 95.** Solve this equation: $x^4 - 4x^2 + 3 = 0$

  Assume that $x^2 = z$ to get $z^2 - 4z + 3 = 0$, which leads to a = 1, b = -4, c = 3

$$z = \frac{4 \pm \sqrt{(-4)^2 - 4(3)}}{2} \text{ which leads to this:}$$

$$z = \frac{4 + \sqrt{16 - 12}}{2} \text{ and } z = \frac{4 - \sqrt{16 - 12}}{2}$$

$$z = \frac{4 + \sqrt{4}}{2} \text{ and } z = \frac{4 - \sqrt{4}}{2}$$

$$z = \frac{4 + 2}{2} \text{ and } z = \frac{4 - 2}{2}$$

$$z = 3 \text{ and } z = 1$$

This leads to $x = \pm\sqrt{3}, x = \pm 1$

- **Problem 96.** Solve this equation. $2x^4 - 7x^2 - 9 = 0$

  By doing the substitution of $x^2 = z$ we get $2z^2 - 7z - 9 = 0$

  Now solve it with a = 2, b = -7, c = -9

$$z = \frac{7 \pm \sqrt{(-7)^2 - 4(2)(-9)}}{2(2)} \text{ which leads to this:}$$

$$z = \frac{7 + \sqrt{49 + 72}}{4} \text{ and } z = \frac{7 - \sqrt{49 + 72}}{4}$$

$$z = \frac{7 + \sqrt{121}}{4} \text{ and } z = \frac{7 - \sqrt{121}}{4}$$

$$z = \frac{7 + 11}{4} \text{ and } z = \frac{7 - 11}{4}$$

$$z = 9/2 \text{ and } z = -1$$

(Again) with the imaginary numbers, you get $x = \pm\frac{3}{\sqrt{2}}, x = \pm i$.

However, final answers do not have a radical in the denominator, so we will rationalize that answer.

$$x = \pm\frac{3}{\sqrt{2}} \cdot \frac{\sqrt{2}}{\sqrt{2}} = \pm\frac{3\sqrt{2}}{2}$$

$$x = \pm\frac{3\sqrt{2}}{2} \text{ and } x = \pm i$$

The answer isn't very simple but you can feel rest assured that mathematicians have an answer for everything, even if it's imaginary.

- **Problem 97.** Solve this equation. $x^4 - 4x^2 - 12 = 0$
  This leads to the "substitution equation" of $z^2 - 4z - 12 = 0$
  This means that a = 1, b = -4, c = -12

  $$z = \frac{4 \pm \sqrt{(-4)^2 - 4(-12)}}{2}$$ which becomes this:

  $$z = \frac{4 + \sqrt{16+48}}{2} \text{ and } z = \frac{4 - \sqrt{16+48}}{2}$$

  $$z = \frac{4 + \sqrt{64}}{2} \text{ and } z = \frac{4 - \sqrt{64}}{2}$$

  $$z = \frac{4+8}{2} \text{ and } z = \frac{4-8}{2}$$

  z = 6 and z = -2

  Then we get $x = \pm\sqrt{6}$ and $x = \pm i\sqrt{2}$

This next three questions involve solving cubic equations, which can be very complex and difficult to do. We'll do some solving of some easy ones so you can see how the math goes with cubic equations. They involve finding the factors, which is something we've looked at in the past but with easier questions.

- **Problem 98.** Solve this equation. $x^3 + 6x^2 + 11x + 6 = 0$. Okay, brace yourself and remember factoring: Start figuring out all the factors of 6 (which means all the numbers that can evenly be divided into 6 and you get plus or minus 1, 2, 3, 6. Now look for a number that makes this polynomial true when substituted for x:

  1: $(1)^3 + 6(1)^2 + 11(1) + 6 = 0$ (this is an untrue statement)
  -1: $(-1)^3 + 6(-1)^2 + 11(-1) + 6 = 0$ (this is a true statement) so x = -1
  2: $2^3 + 6(2)^2 + 11(2) + 6 = 0$ (this is an untrue statement)
  -2: $(-2)^3 + 6(-2)^2 + 11(-2) + 6 = 0$ (this is a true statement) so x = -2
  3: $(3)^3 + 6(3)^2 + 11(3) + 6 = 0$ (this is an untrue statement)
  -3: $(-3)^3 + 6(-3)^2 + 11(-3) + 6 = 0$ (this is true) so x = -3
  6: $6^3 + 6(6)^2 + 11(6) + 6 = 0$ (this is an untrue statement)
  -6: $-6^3 + 6(-6)^2 + 11(-6) + 6 = 0$ (this is an untrue statement)
  This leads to the factored equation of: (x + 1)(x + 2)(x + 3) = 0
  x = -1, x = -2, x = -3

- **Problem 99.** Solve this equation. $x^3 - 2x^2 - x + 2 = 0$

  There are only four factors to consider in this one because the last number is 2: -1, 1, -2, 2

  So, let's try this again:

$$-1: (-1)^3 - 2(-1)^2 - (-1) + 2 = 0 \text{ (this is true) so } x = -1$$
$$1: 1^3 - 2(1)^2 - 1 + 2 = 0 \text{ (this is true) so } x = 1$$
$$-2: (-2)^3 - 2(-2)^2 - (-2) + 2 = 0 \text{ (this is false)}$$
$$2: 2^3 - 2(2)^2 - 2 + 2 = 0 \text{ (this is true) so } x = 2$$

  This means that $(x - 1)(x + 1)(x - 2)$ is how this factors out.

  The solution is: $x = -1, x = 1, x = 2$

- **Problem 100.** Solve this equation. $x^3 - 4x^2 - 7x + 10 = 0$

  We'll do some creative factoring with this one: The factors of 10 are $\pm 1, \pm 2, \pm 5, \pm 10$.

  Start with 1 and see what happens: $1^3 - 4(1)^2 - 7(1) + 10 = 0$. This is true! It means that the factor of $x - 1$ is one of them. Now you can "creatively" divide $x - 1$ into it by clumping groups that can be divided nicely into $x - 1$:

$$(x^3 - x^2) \, (- 3x^2 + 3x) \, (- 10x + 10) \text{ leaves you with this:}$$
$$(x - 1)(x^2) - 3x(x - 1) - 10(x - 1) = 0 \text{ Divide the entire thing by } x - 1$$

  to get this:

$$x^2 - 3x - 10 = 0$$

  Now you can solve this with a = 1, b = -3, c = -10

$$x = \frac{3 \pm \sqrt{(-3)^2 - 4(-10)}}{2} \text{ which leads to this:}$$
$$x = \frac{3 + \sqrt{9+40}}{2} \text{ and } x = \frac{3 - \sqrt{9+40}}{2}$$
$$x = \frac{3 + \sqrt{49}}{2} \text{ and } x = \frac{3 - \sqrt{49}}{2}$$
$$x = \frac{10}{2} \text{ and } x = \frac{-4}{2}$$

  x= 5, x = -2 and don't forget that x = 1 as well.

  I don't know about you, but I thought that was the hardest question yet.

So, basically, we've traversed our way through the easy, the medium, and the seemingly impossible algebra questions. If you're not a mathematician, you may be able to forget some of what you've just learned as they don't apply to real life. Even so, there were plenty of word problems to indicate that, in fact, algebra can be used to solve problems that you might encounter in real-life circumstances. We hope you enjoyed this guide!

If so, can you leave a review on the Amazon book page? It would be greatly appreciated!

If you have any suggestions on ways to improve this book. Please contact us at: support@mathwizo.com

Did You Get My Free Guide?

For a limited time, you can download this book for FREE!
Get it by going to: https://go.mathwizo.com/1

Made in the
USA
Monee, IL